I0233244

The Intelligent Creation of Life

Adam Mann

Clink
Street

London | New York

Published by Clink Street Publishing 2015

Copyright © 2015

First edition.

The author asserts the moral right under the Copyright, Designs and Patents Act 1988 to be identified as the author of this work.

All rights reserved. No part of this publication may be reproduced, stored in a retrieval system or transmitted, in any form or by any means without the prior consent of the author, nor be otherwise circulated in any form of binding or cover other than that with which it is published and without a similar condition being imposed on the subsequent purchaser.

ISBN: 978-1-910782-22-4
E-Book: 978-1-910782-23-1

Table of Contents

1. Introduction

A good way to start perhaps is to explain that you, the reader, are literally a star child. As you will see shortly every part of you, absolutely literally, every single component of every atom in your body came into existence at the moment the Universe itself came into existence.

You really are a Star Child.

It was when time itself started.

This book is going to take you on an amazing voyage. It is going to enable you to understand the way in which our planet Earth, the Solar System and our Sun, around which the planets orbit, formed. ou will follow the development of life, from First Life through to the human beings we are today.

The material we are considering here discusses the very basic factors underlying the initial occurrence of the First Life and its progressive development to the human beings we are. The book has been deliberately designed to be relatively short and concise, with the discussion of events in each chapter seeking to provide a brief but clear explanation accessible to any reader. Mathematics and unnecessary complexities have been studiously avoided.

As you progress through the book you will see a series of sequential events unfolding. Could these events, either individually or as a collection, possibly have been due to a series of random events? Alternatively are they the consequence of an Intelligent Design process occasioned by an Intelligent Creator

?

The Big Bang, the start of everything including time.

The occurrence of the Universe coming into existence is known as the Big Bang, an event which occurred some 14,000 million years ago. It marked the beginning of everything we know, including the very concept of time itself. We discuss the nature of the Universe shortly. For now suffice it to say that everything on Earth and in the skies about us exists as a result of that occurrence.

Our precious Solar System is in our part of the Universe. It consists primarily of the Sun, The Earth, our Moon, seven other planets and their moons. The Solar System developed through a series of clearly defined events that led to our Earth

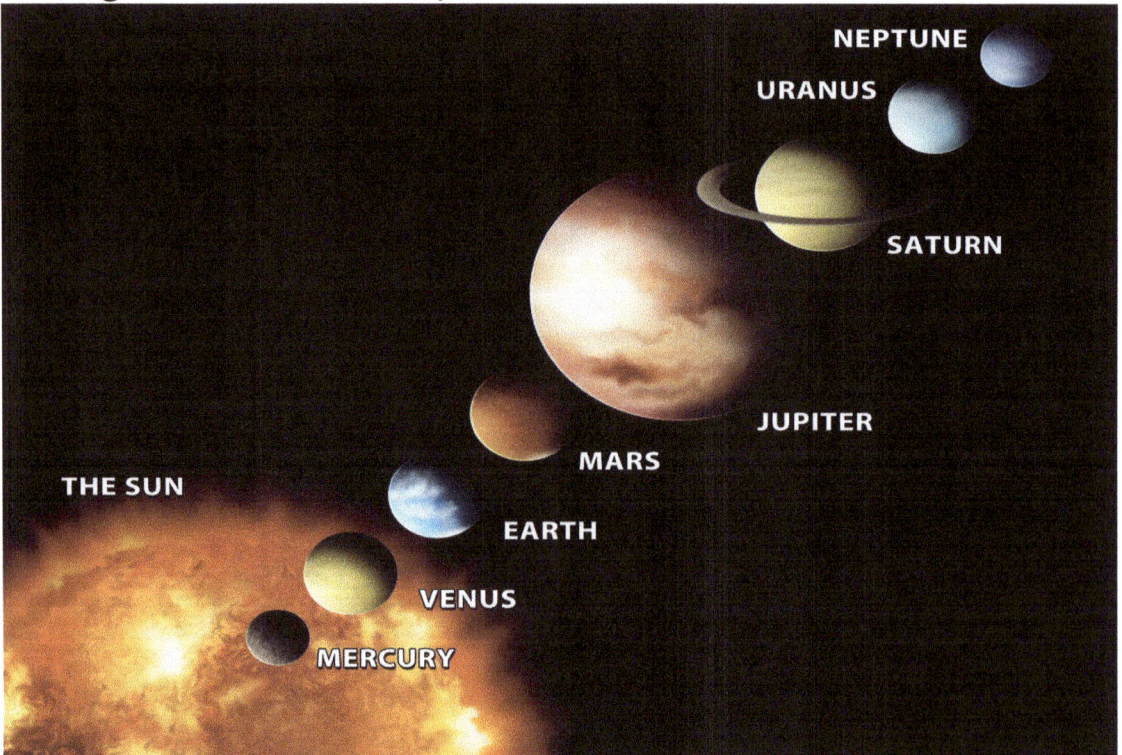

This is the Sun, the Earth, third out from the Sun, and the seven other orbiting planets. Pluto was also designated a planet, but was reclassified as a dwarf planet in 2006.

The reference to "our part" of the Universe in the earlier paragraph was a deliberate one. The Universe is unimaginably vast, and the space we occupy in it, the Solar System, relatively tiny.

We humans can look out into the Universe but we are physically constrained to our Solar System by the many, many millions of miles of space between us and our closest neighbour, Alpha Centauri. Alpha Centauri is the closest star system to us at a distance of 40,000,000 million kilometres. [I have used this form of number 40,000,000 million, rather than a 40 trillion because it hopefully gives a more objective concept of numbers.] To the naked eye Alpha Centauri looks like one star, but it is actually a three star system.

We can look out into the Universe, but distances to our neighbours outside the Solar System are unimaginably vast.

We are increasingly able to physically explore the Solar System by the use of manned and unmanned expeditions. A manned landing on the planet Mars is a realistic possibility. Our knowledge about the rest of the Universe is, however, limited. We can observe the Universe through a range of optical and other detection devices and make assumptions about its nature. It is in this way we have been able to establish that Alpha Centauri comprises three stars, not one.

Being able to discover that a point of light is made up of three stars is a significant technological achievement. Knowing this does not, however, tell us much about the system, and that's our closest neighbour! We have recently discovered that some 95% of the Universe, appears to constitute a combination of "Dark Matter" and "Dark Energy". It may seem unbelievable but we literally have absolutely no idea what "Dark Matter" and "Dark Energy" are. Basically, whilst may we conjecture, we know so very little.

Returning to the Solar System, an analogy of the relationship between the Solar System and the general Universe is perhaps of a young child in the back yard of its parents' house. The child is able to examine what is in the back yard and come to conclusions. Some of these may be correct and some may not.

The child then peeks through the fence at the outside world. The child has absolutely no tangible way of even starting realistically to think about and explore what is out there. It will be a mysterious realm where the child may, for example, see a car. The car will just be there, yet the child will have no concept of the factories where cars are made, workforces and so very much else.

A child looking out may see all that an adult can, but lacks the ability to understand and explore.

We can then, perhaps, think of the Solar System as the young, fledgling human race's back yard comparable to that of the child's. We are increasingly able physically to explore it and understand its nature. The concept of the Solar System being "our part" of the bigger Universe is a perfectly reasonable one. When we look out at the Universe, however, we are as ill-equipped as the child in its garden to understand what underlies it!

In this context it's curious to reflect on, just as a starting point, the actual nature of the Solar System. Each of the planets is very different, as are the one hundred and fifty plus moons that orbit them. Then there are myriad comets and asteroids.

If part of the design of the Solar System was to give us an interesting backyard to explore, then that design certainly succeeded.

Less than 100 years ago H. G. Wells published the then ground-breaking book "The Outline of History" with over two million copies published. It was described then as "the whole story of life and mankind" (sic), humankind (descriptive gender wasn't an issue in those days). The contrast between what was known then, in comparison with now, is stark. H. G. Wells's book talks about the concept of life, the Universe and the Solar System in passing, in just a few pages.

There is no mention in the book of DNA or genetics. The depth of history of our existence, nature and development then is scant. This is not intended in any way as a criticism of the book. It sold two million copies. It is simply the fact that just that very short time ago, almost nothing was known about the true nature of our wold and our development in it. The information was simply unknown, waiting in the background to be discovered.

Advances in modern science in the short space of time since then have been incredible. Our understanding of the Earth and the human beings on it is such that we have developed a significant depth of understanding of our origins. A word of caution here, however. That having been said, as you progress through the book you may very well come to consider the extent to which what we "know" is heavily outweighed by what we don't know.

One way that science can be described "technically" is as an activity that is intellectual and practical. It involves the objective study of how the natural world functions. A particular feature of science is that its nature is defined by particular rules. In particular science requires not only to be able to observe phenomena, but also to test the results of observations.

Because science is unable to test the concept that life is the consequence of Intelligent Creation it cannot, within its current approach, consider life in that context. Science thus looks to any explanation that can possibly avoid the concept of Intelligent Design and Intelligent Creation. In so far as there are no tangible explanations, science resorts to the stock concept "we don't yet understand [but we may somehow in due course is implicit]".As we will see later there is indeed a test, a statistical one, that can be applied to the concepts considered here.

The general scientific approach to the occurrence of life is that it is a consequence of a series of random events. The same applies to the existence of the Earth and the later development of humans on it. It follows a concept of determinism. Determinism is the physiological doctrine that every state of affairs, including every human event, act, and decision, is the inevitable consequence of a previous, ongoing chain of events.

Determinism holds that the human race exists simply as a consequence of the way things are. It is an argument that holds that life on Earth is simply the logical outcome of the progressive series of events that led to it. Determinism holds that for every possible event, there exist conditions that could cause no other events.

This historical ethos of science's absolute need to test is understandable. It was believed, perhaps hoped that it was possible to achieve and understand a knowledge of everything about us. Concepts then were however "rudimentary".

Democritus for example developed the idea of atoms. If

you break a piece of matter in half, and then break it in half again, how many breaks will you have to make before you can break it no further? He thought that it ended at some point, a smallest possible bit of matter. He called these basic matter particles, atoms.

Today, in this modern age we are able to pursue research more and more deeply. Discoveries make it more and more clear how incredibly complex the world around us is. Whilst there is much we still don't understand, we have learnt much about the course of developments of life over time. We also have started to understand the mechanisms that drive it.

What we do not know is how life originated and, in many respects, developed. For science simply to refuse to even consider the concept of Intelligent Creation, driven by Intelligent Design, denies it an opportunity to widen the depth of its exploration.

A major breakthrough in being able to understand the mechanisms of life was the understanding of the basic nature of DNA (deoxyribonucleic acid, if you feel a need to know the full title). DNA is the primary matter that controls how offspring inherit their characteristics from their parents. DNA also controls every aspect of the development of an organism from birth, and maintains and repairs the organism throughout its life.

DNA consists of two long chains of molecules twisted into a double helix. The double helix is linked by a series of bonds that hold the chains temporarily together. When a particular section of DNA needs to perform a function, the bonds at the specific point where the necessary DNA is located unwind to enable that particular section of DNA to perform its function.

Human DNA contains some 6,000 million basic components (nucleotides if you want to do an internet search), made up from some 200,000 million atoms.

DNA is extremely complex and there is much about it that we need to learn and understand. We do know however that it can be equated to a complex computer code of instructions, although there are many instructions that we have been unable to interpret. DNA is a language understood by every living organism on Earth.

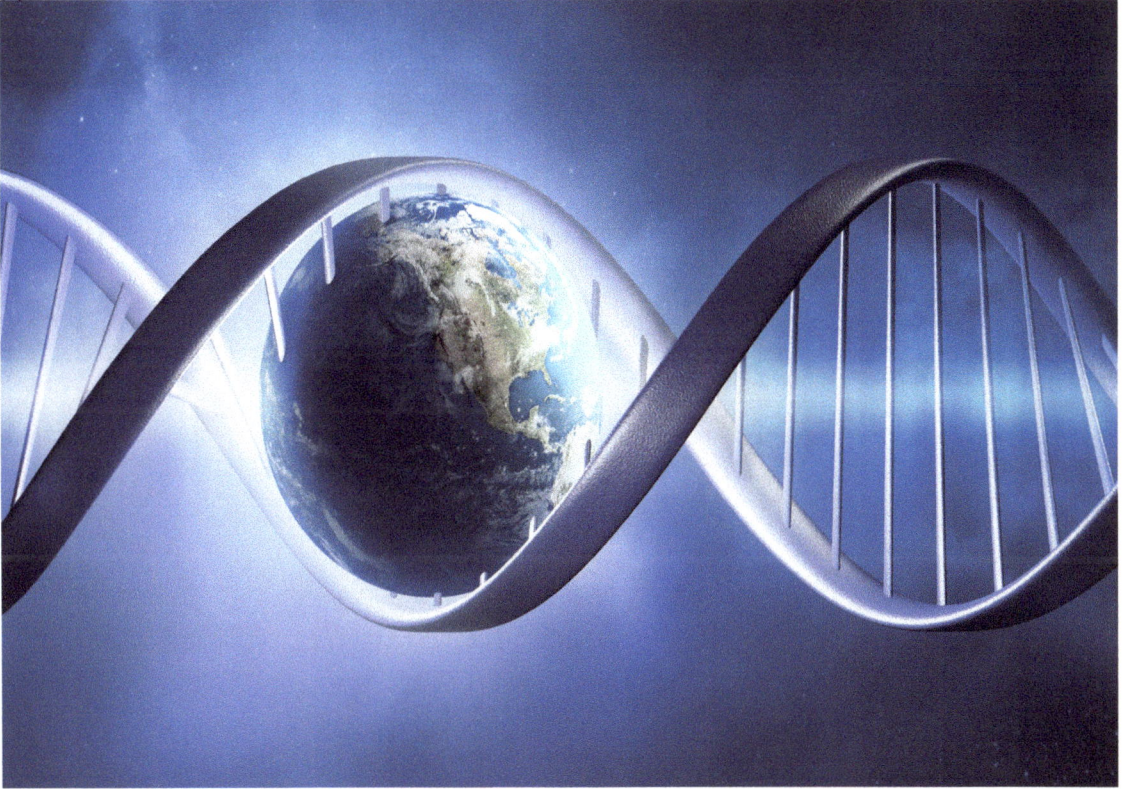

It is an incredible fact that every life form on Earth understands the same DNA language and can act on its instructions.

By reference to the previously considered deterministic concept, our first endeavour is to explore the coming into being of DNA and First Life. How likely is it that such a complex code, along with the occurrence of the First Life, could have occurred randomly?

2. The Universe and a Starting Point

Before we can move to consider the underlying nature of life we should first consider the very nature of the Universe. The very first challenge to determinism, although not a direct one, is the nature of the Universe. In this chapter we explore the basics of what happened in the instant after the Universe came into being.

The reference to the "instant after the Universe came into being" is important. It creates a distinction between the nature of events that occurred before the coming into existence of the Universe, and the events that occurred immediately thereafter. This book does not in any way consider events that occurred prior to the Universe coming to existence.

Elements are atoms, basic chemical substances that cannot be broken down into something smaller using chemical methods. Literally everything is made from atoms. Atoms are tiny, tiny structures made from a central nucleus containing protons and orbiting electrons. Hydrogen is the most basic of atoms, comprising one orbiting electron and one proton in its nucleus.

Hydrogen is the building block for heavier atoms which are formed when additional protons and electrons come together to form larger atoms. Generally the number of electrons and protons defines the atom and the element represented by it. The element carbon for example has six protons, the element oxygen eight and the heavier element gold seventy nine.

As noted earlier it is believed that the Universe came into existence some 14,000 million years ago in an event known as the Big Bang. At this time the basic light atoms of hydrogen formed virtually spontaneously. These hydrogen atoms were pulled together by forces including gravity to form into large clouds of dust and gas.

These vast clouds are known as nebulae and were the basis for the formation of the first stars. As the hydrogen atoms were crushed tighter and tighter together through forces such as gravity, individual hydrogen atoms joined together to form heavier atoms. Helium was formed from two hydrogen atoms crushed together, lithium from three and carbon from six.

There are ninety naturally occurring elements. The elements are numbered from 1 – 92, increasing in weight and complexity as the numbers increase. (43 and 65 do not occur naturally so ninety is a valid number for our purposes). Elements up to and including iron (26) are manufactured in "ordinary" stars such as our Sun by gravitational and other processes just described.

Once the star gets to the point of having manufactured iron, the processes by which the elements are combined in an ordinary star are at an end. None of the heavier elements can be constructed by this mechanism of being crushed together by the ordinary process in stars.

The nature of the Universe is such that it required the remaining sixty four elements to make the full set of ninety required for human existence as we know it There is a specific process built into the nature of the Universe by which those sixty four additional elements are manufactured.

The process is not confined just to a manufacturing one,

but also a secondary process of the distribution of all the elements. As you will see the process is such that the full set of necessary elements was available for the formation of the planets such as Earth, and the life thereon.

This formation and distribution of the heavier elements occurs in a process known as a supernova.

A supernova is a star that emits a vast amount of light as it explodes. A supernova seen in 185 A.D. remained illuminated in the sky for some eight months.

This is an extremely powerful stellar explosion that is so violent that it can literally, for a time, outshine an entire system of 1,000s of millions of stars known as galaxies. During a supernova explosion the dying star, in a relatively short space of time, may emit as much energy as our Sun will have emitted during its lifetime.

During a supernova the remaining sixty four elements are

created, and all of ninety elements are then expelled into space in the massive explosion. This is exactly what occurred in our own Solar System some 5,000 million years ago when a predecessor to our Sun went supernova.

The exact nature of the formation of the Solar System is not known for certain. It is believed that the supernova interacted with an existing gas and dust cloud present at the time causing it to form into a revolving disk which speeded up. As the speed of revolution increased, more and more mass was drawn into the centre. A build-up of heat at the centre of the disk may have formed what is known as a protostar, essentially the collection of gas needed for star formation. The pressure of the gas would have increased until nuclear fusion started and the Sun as we know it today came into existence.

Not all the material from the supernova was absorbed by the sun. A cloud of debris remained and this debris coalesced, forming larger and larger particles. These particles formed rocks, and as larger and larger rocks collided they were held together by gravity. Individual collections of rocks grew in size, leading to the formation of the planets including of course the Earth.

When life subsequently formed on the Earth, the elements that make up every organism on it were those emitted from the supernova. The materials from which those elements are made had come into existence at the time of the Universe coming into existence. You were previously described as a Star Child. You now know why, and that every atom in your body has passed through a sun and been part of the blazing glory of a supernova.

Incidentally, different combinations of those ninety elements make up literally all the different types of known materials. Our bodies, the sea, and the atmosphere, for example, are all made up of different combinations of elements. An analogy of the different combinations of elements is how the twenty six letters in the English alphabet make up all the different words in the English language.

Following the supernova and the formation of The Solar System with the Earth within it, the nature of the Earth and its location was ideal for the Intelligent Creation of life.

We have not considered at this stage the detailed formation of the Solar System and the Earth therein. Nonetheless, the factors that influenced the very nature of the Earth, and its habitable orbit in the Solar System were critical even to the ability of life to occur. A consideration of this is included in Annex 1.

3. Before First Life

We are now at the point where we can get to a consideration of specifically identifiable occurrences that may have led to the occurrence of life. The detail and the consequences can be tangibly considered.

The early Earth formed some 4,500 million years ago. It was a bubbling ball of molten rock heated by the original supernova and ongoing collisions with rocks. After its formation the Earth continued to be bombarded by meteors and asteroids and cooled only as the bombardment lessened.

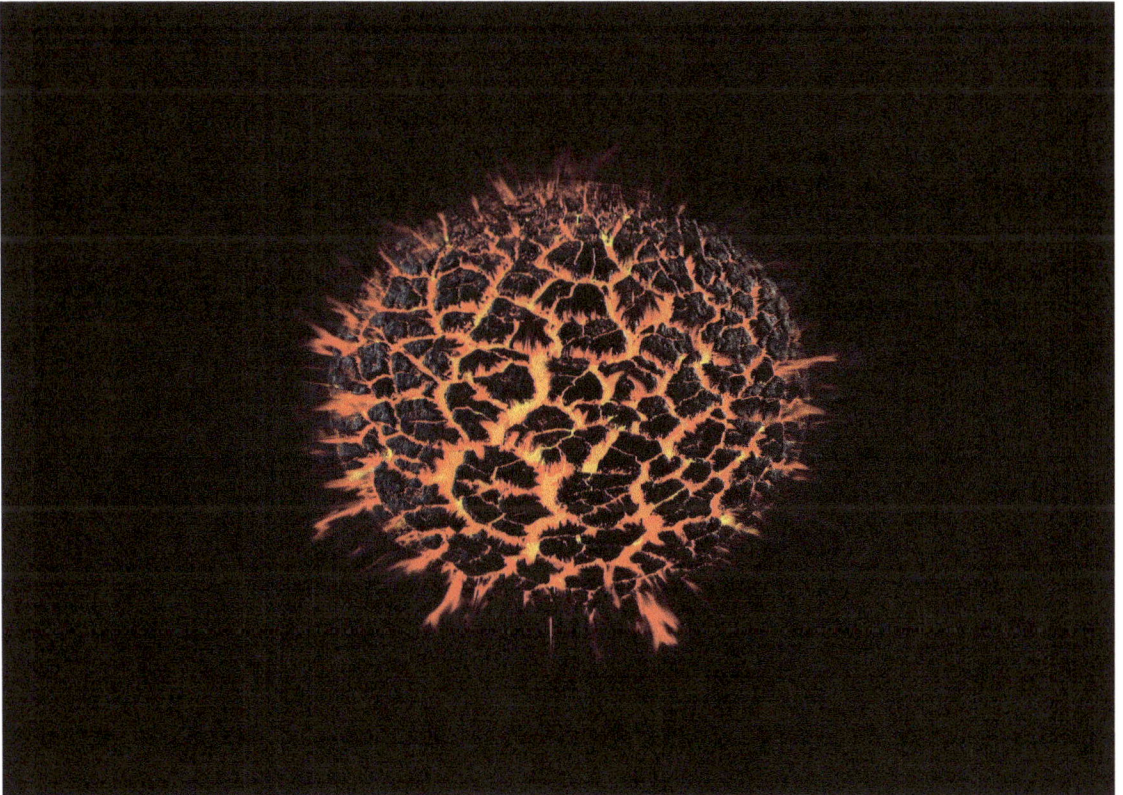

The early Earth was a red hot ball of molten rock for a long time.

Many of the asteroids bombarding the Earth carried water in the form of ice which evaporated during the original bombardment. Over time, the bombardment lessened and the planet cooled. Water-carrying asteroids started to be able to retain the moisture instead of it being evaporated away. Water started to be present on earth and to increase. The water in turn condensed into the atmosphere and rain, along with more water-bearing asteroids, caused the flooding of the seas. Water, so very precious and necessary for life, was delivered to the Earth.

The nature of the Earth became such that it has been able to retain that water. The volume of water was, interestingly, of the "correct" amount so that it did not cover the whole planet. Had the planet become covered in water, and devoid of dry land human life could not have occurred.

Asteroids delivered water of the volume currently present to the Earth.

As the early Earth cooled and the asteroid bombardment had settled, a thin crust formed on the surface of the World. It was still, however, a very hostile place by our standards. The atmosphere was devoid of oxygen, acidic and would have been extremely toxic to us. Volcanoes abounded.

Volcanoes abounded on the Earth.

The Sun was dim. The Moon, which had formed earlier on in the Earth's development, was much closer than now and loomed large on the horizon.

The gravitational attraction of the Moon is the primary cause of Earth's tides. As it was then much closer to the Earth than now its gravitational effect would have been a contributing factor in the violent tides that existed then.

Violent storms raged across the early Earth.

Amid all this large scale activity, much was happening on early Earth at an atomic, chemical level. Now is a good time discuss the concept of chemistry, in context, without getting into too much complexity. We considered earlier the basic concept of atoms, their structure and nature in the context of specific elements.

We can now usefully look at elements in the context of the human body. Almost 99% of the human body is made up of six elements: oxygen, carbon, hydrogen, nitrogen, calcium, and phosphorus. There are others, but these are the main elements.

These six elements are not static in their nature but are capable of, and undertake, interactions with other elements. Many millions of chemical interactions occur continuously as your body functions. You take in food, the food is broken down into the elements and those elements are then reassembled into molecules.

The word molecule is simply the descriptive term for a particular assembly of atoms. You may have seen a reference to

H_2O in relation to water for example. This is simply a statement in chemical terms that each water molecule is made up of 2 hydrogen atoms (H_2) and one oxygen atom (O). There are many different types of molecules, and many different types of chemical reactions.

Let us now move to understand more about life itself, focusing on DNA. We looked at the basic structure of DNA earlier, but what is it? What do those 200,000 million atoms do? The image shows a DNA sequence where each of the coloured rectangles represents a part of a specific set of instructions.

Although not written in computer binary code the functionality of DNA is similar.

A very important point here is that the very genetic code that controls your body is the same genetic code that is lodged in every life form on Earth. I mean that literally. Place a piece of DNA from an elephant, with a particular set of instructions, into a blade of grass and the blade of grass's DNA could read that code. Of course if the instructions of the code was to create a tusk that blade of grass would not have the necessary ingredients

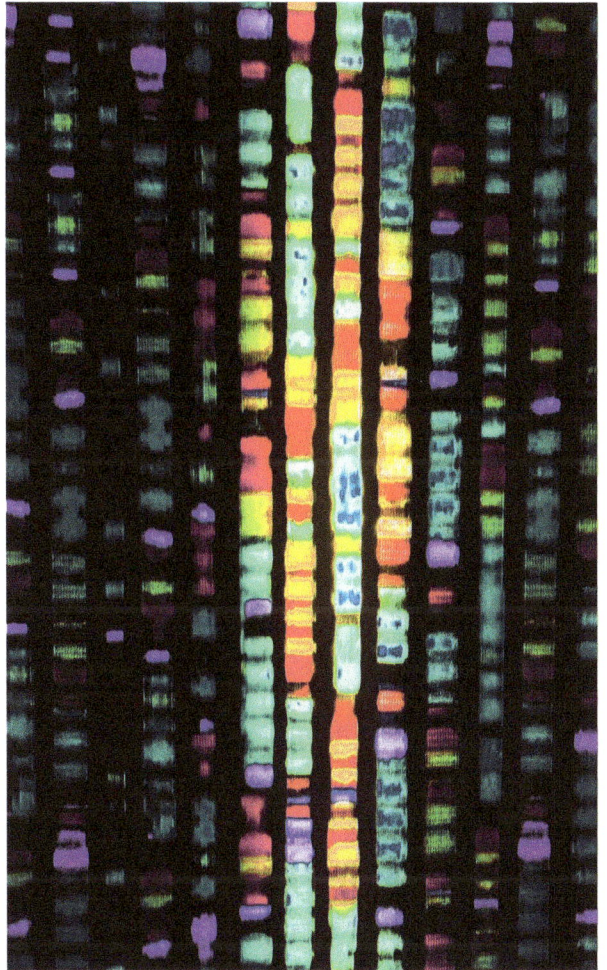

to make the tusk, although it might give it a try.

This is what genetic engineering is all about. Inserting DNA from one organism into another to bring about changes in that organism. The elephant example was a deliberate extreme to make a point. The transfer of DNA with a particular feature from one plant into another plant to cause the recipient plant to acquire that feature is what genetic modification of crops is all about. Genetic engineering here is used to alter flavour, shape, and texture.

However, getting back to our main theme, life first developed on Earth some 3,500 million years ago or even longer. What is described as the Last Universal Common Ancestor (LUCA) is an organism, a single cell that first came into existence at that time. Incredibly it has been established that that single organism contained the same DNA code as every living life form on Earth today. The DNA code that exists today in your body is identical to the DNA code that was in the first known living life form.

So then what is life, how is it defined? Let us cut to the quick, or at best, let us stumble clumsily across an attempt to define what life is. It is a good question indeed. Take a multi-mineral tablet and a multi-vitamin tablet. Crumble them together into a tumbler of water, add some energy and you have most of the ingredients of life. How did such a mixture of basic elements come together to form life? Despite massive research and experimentation, science has absolutely no idea whatsoever.

It gets worse. Scientists across the world have been unable even to agree on a description that embraces all aspects of life, in order to adequately define it. One definition could be that "life is defined as something possessing the ability to grow and reproduce, and function". There are many alternative versions.

As we shall see later there is a very good reason for this perplexed situation as we seek to quantify just what it is that defines humans. No single description of life can encompass a description that includes both human and non-human life. The development of human life in the last 12,000 years has been such as to render it unique amongst life on Earth.

As you will see there can be no common definition of the two!

4. First Life

We start this chapter with some thoughts about amino acids. You may well have seen excited reports citing the possible location of non-terrestrial amino acids, their formation and even their arrival on meteorites from space. Given the amount written about amino acids, before we press on it is worth addressing this amino acid issue.

The presence of amino acids does not in any way confirm, or even realistically imply the existence of life at any particular location.

Taurine

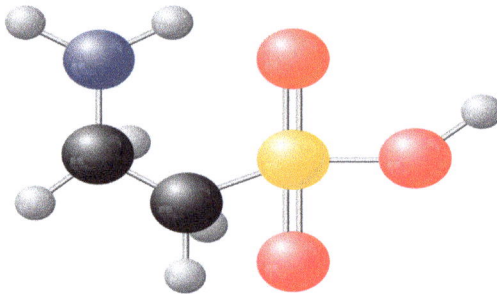

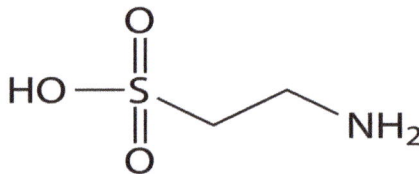

Taurine is a widely distributed amino acid in animal tissues and plays an important role in cardiovascular function.

Amino acids, of which taurine is one, are specific combinations of atoms of which there are twenty different types in the human body. They convey clearly defined chemical information and, like the letters in the English alphabet, can be arranged in an infinite number of ways to convey that information.

Science has long strived to find explanations for life's existence.

Miller-Urey

Abiogenesis is the theory of the spontaneous development of life from non-living matter. It suggests that somehow, some chemicals can randomly come together and, with the inclusion of energy, lightening for example, successfully combine and become alive. An experiment known as the Miller-Urey experiment in 1952 tried to recreate the then-believed conditions of the primitive Earth. It succeeded in synthesizing organic molecules, but not life.

Countless attempts occurred both before Miller-Urey and subsequently since then to try to bring life into existence from inanimate material. None have succeeded.

Charles Darwin

Charles Darwin's theory of evolution is frequently muddled into the concept that it is trying to address the development of early life. It isn't. Rather it suggests that all life on Earth has descended from some primordial form into which life was first breathed, which is an entirely different thing.

Many of Charles Darwin's achievements are attributed to his discoveries on the Galapagos Island. This Galapagos Tortoise was just one of them.

Charles Darwin's theory suggests that a biological change that is significant is first occasioned by variations that arise randomly. These variations must be subsequently heritable and survive competition for survival. The process in essence is one of a sequential improvement of existing life forms by natural selection. The emphasis is on natural selection of existing life forms leading to the improvement of those life forms.

Natural selection provides that the best of a species survive, and those best go on to produce better. Species thus vary their nature and form to develop in relation to changes in their environment and circumstances. New species develop from existing ones by changes to already existing features.

When Charles Darwin announced his theory, science's

knowledge was at a very different level from today. As our knowledge has increased, this theory has increasingly, in a broad sense, been interpolated to embrace newer discoveries about life's early development.

Even though the theory of Natural Selection did not specifically address the origin of life, effectively it is frequently confused into the concept that it does so. It does not. As we discuss the origin of life it is important to be clear that the coming into existence of the First Life is not attributable to natural selection. Additionally the ongoing development of new and radical life forms that we shall go on to discuss is frequently and incorrectly attributed to the theory.

That an organism changes and adapts its existing form, the emphasis being on "existing", to improve the existing form, in accordance with the concept of the survival of the fittest, is completely rational. It is clearly seen on a day-to-day basis and can properly be attributed even to changes from one species to another, with similar characteristics.

Concepts of Intelligent Creation

So what is suggested by the concept of an Intelligent Creator, responsible for the Intelligent Design of life, leading to the coming into being of the human race? The previous consideration of a young child in the back garden was a salient and deliberate one.

The child, with all its human features including intelligence, simply is not sufficiently developed to be able to understand the outside adult world. The child cannot even understand the house and the back yard; who built it and for what underlying purpose? As the child grows and develops it will learn and understand these things. Was the development of the Earth and the occurrence of life of a similar ilk? Is there some Intelligent Creator, and a bigger concept which the human race may be enabled to be able to understand in time?

In relation to us now can we be compared to a nest of ants? There are many similar aspects to a human colony. The ability to communicate, communal living, warfare, kidnap, nurseries for the young and many more. The nature of the ant, however, will never be able to make any sense of the bigger world around it.

Are we humans constructed in a similar fashion relative to the Universe and an Intelligent Creator? Is there, in front of us and all about us, some sort of existence that is invisible to us. As the ant cannot perceive of our living world is it that we simply aren't endowed with the ability to see this bigger existence?

Interestingly science is increasingly moving to a recognition and acceptance of this concept of the unknown world about us that we simply cannot perceive. We have touched already on unknown dark matter and dark energy. Another increasingly popular concept is "String Theory". This theory suggests, in essence that there is no such thing as matter as a tangible substance. Rather everything, atoms themselves consist of literally energy and we live in a universe with multiple dimensions.

We have considered just a tiny few of the vast numbers of matters that attest to the structured, ordered and complex nature of the Universe and the basic elements of life. We can now go on to explore the equally, or perhaps more, complex matters associated with the occurrence and development of life.

It is perfectly logical and rational to incorporate the concept of an Intelligent Creator who has actually planned the Intelligent Creation into a consideration of the occurrence and development of life. Whilst it is not possible even to start thinking about how this was done, we can properly consider the processes that might have been involved.

If we accept, as we reasonably can, that the process has occurred in a logical fashion then it is equally as logical to propose that this is a consequence of a logical plan. On that basis it is equally as reasonable to go on and consider that the concept of natural selection was not the keystone for the development of life.

Rather was Natural selection (just) one part of an overall plan underlying the development of life once it had started. If we are prepared to consider that the occurrence of life did not occur randomly, then we can equally as reasonably consider that some form of Intelligent Intervention into the Earth's development regime occurred.

We need then to consider what would be involved in a random selection process

An Instantaneous Random Assembly of Molecules

As is the case today, vast numbers of chemical reactions were occurring on the early Earth. Two possible models underlying the random assembly concept are that of instantaneous self-assembly, or, alternatively, staged random assembly. We consider first the instantaneous concept.

This concept must have it that numbers of molecules spontaneously organised themselves into a one-off living life form, the first cell. There was no natural selection here, that didn't kick in until after the first life form existed. Let us think about what would have to occur in the event that this concept of random self-assembly of chemicals giving rise to life could successfully have actually happened.

We have seen that DNA and the cell itself are two distinct mutually-dependent entities. The cell is, if you like, the hardware that houses the DNA and the DNA is the software that drives and directs the cell. By way of a comparison of complexity of development, think about a complex laptop computer and the underlying software that drives it. This concept of development of life is of an entirely different, greater scale, in terms of complexity than the laptop, no matter how modern it is.

When thinking about the physical computer, don't think just about the hard drive. Think about all the components of the hardware that make it a functioning item. Not just even the myriad other electronic components but the keyboard, screen and mouse. Think also about the complexities associated with the very concept of processing data. Then think separately about the operating software that drives the functionality of the computer and the software programs that are functioned by the computer to produce output. Forget about human input.

Then consider a computer and its software as another analogy. They must interact together in complete harmony and have been developed simultaneously, but separately. Not only must their development be simultaneous but it must also be congruent; the software and the hardware must be mutually compatible.

Now look at the concept of First Life having been randomly developed, mirrored by the concept of a computer being randomly developed. Yes, they require different facilities and processes, but it's just the principle of complexity that is explored here.

The First Life had in essence the same concepts of components, hardware and software as a computer system. It was, however, far more complex than a computer. Additionally the First Life had to understand and develop many completely different concepts, survival, reproduction, why it should reproduce, and many more. In these respects it was instantly way ahead of a computer.

Once the First Life had reproduced, its successors had to inherit that ability and learn somehow predation and defence. And so it goes, on and on.

Finally and more importantly it had to become alive. Not only can we not figure out how the First Life became alive, but we do not even know what life is to the extent that science cannot even now agree on a description of life.

A fundamental problem with the random concept of the coming into existence of the First Life is that there is no development program of improved design, no prototyping and improvement of designs as a process. As noted, the equivalent of Darwinian evolution per se doesn't cut in until the first unit is in place. In the event of this totally random process every single atom of every single molecule in the first cell would have had to self-assemble randomly.

A Random Assembly by a Staging Process of Molecular Development

OK, so as an alternative to a just one off random self-assembly, suppose somehow that there was some sort of random self-assembly staging process. Groups of molecules formed and developed into more complex basic molecules as a result just of the way things are. The idea has been suggested, it's a popular theory.so let's give it a try.

Perhaps certain types of molecules (previously discussed amino acids, say) did self-assemble into defined groups. These defined groups then further self-assembled into more complex groups. Prototypes must have come and gone; various versions came into existence until the self-assembled process was perfected. Our LUCA randomly came into existence.

But then why just one version of the LUCA with one totally unique set of DNA to be passed on? Where, then, are all the counterparts to the one LUCA? A complex developmental process must inevitably have produced counterparts. They would have been less or differently developed functional versions and equally developed but slightly different versions. Identical versions with slightly differently developed genetic code and so on. Where are they?

If there was no deliberate Intelligent Creation program, how could there possibly have been just one LUCA with an identical genetic code inherited uniquely by every one of the millions of different life forms on Earth over the past 3,500 million years?

Interlude (1)

We looked first at the basic features underlying the existence of life. You will draw your own conclusions as to whether life could possibly have occurred spontaneously. If the whole concept of the spontaneous occurence of life is just a step too far, then what is the only logical alternative? Is it to conclude that the coming into existence of the First Life occurred by some other method, some form of deliberate action, an Intelligent Design by an Intelligent Creator?

5. The First Single Cells

We are now going to take considerations further and explore life more deeply. In doing so we are going to continue our discussion of living organisms without yet having considered and tried to understand the definition of life. Please bear with this and go with the flow and we will come to discuss the definition(s) soon.

Leaving aside scientific complexities of descriptions, perhaps surprisingly there are just two types of cell on the Earth. They are called prokaryotes and eukaryotes. We shall discuss prokaryotes first.

Prokaryotes "Simple" Single
This is the first form of life to have come into existence, as described previously. It is known as a prokaryotic cell, a prokaryote. Prokaryotes are cells that do not have a nucleus in which their DNA is contained. The DNA in it simply floats in the interior of the cell. They do not have any other internal organised structures.

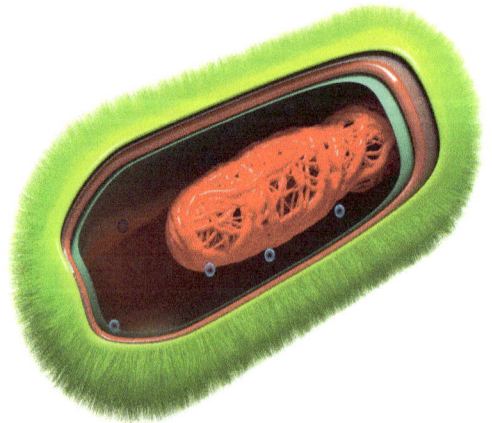

Prokaryotes lack an internal structure and their DNA simply floats within the cell.

Having come into existence, the first prokaryote decided unilaterally, there being no other cells to consult, to engage in reproduction. This first form of reproduction is known as binary fission. It required that first cell to duplicate its DNA code and other cellular components and then divide itself into two new cells, each containing a share of the original / reproduced components.

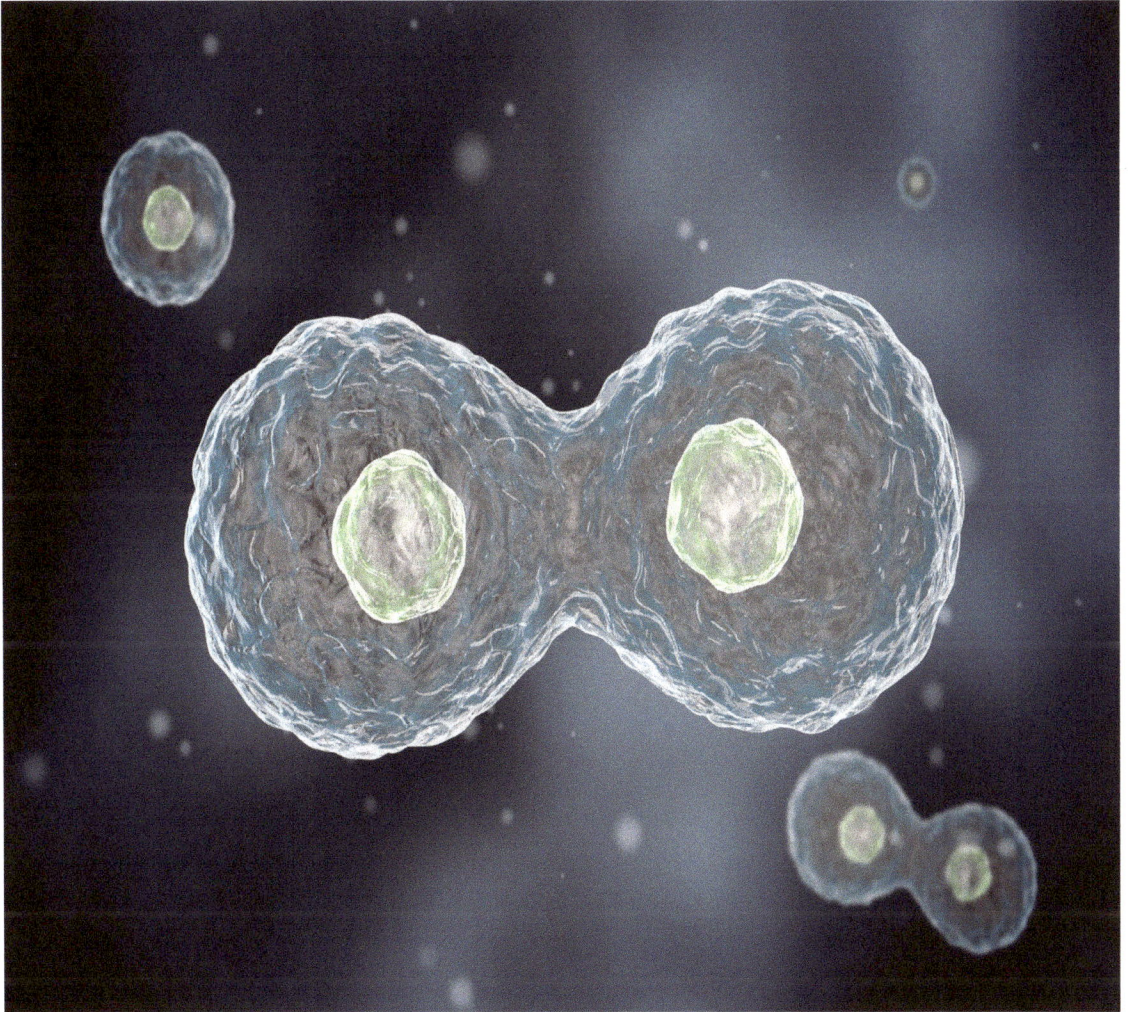

Binary fission involves a process of reproduction whereby a cell divides into two separate identical cells.

Now why would a randomly assembled collection of molecules decide to reproduce? It is difficult to see how it could have been any sort of random process. Although life cannot be adequately described one of the "accepted" definitions of life is the ability to reproduce. Interestingly, just one of many more complexities is that this concept of being able to reproduce itself must have been built into the first cell's DNA before a first reproduction had ever occurred. The intention of reproduction must thus have been a part of the cell's development, an anticipated need.

Life had come into existence relatively quickly after the Earth first formed. The photosynthesis function (explained shortly) that was necessary to clear the atmosphere became part of that life form. The production of oxygen by the new life form was a long process, during which time there was no need for anything else to happen – and little else did.

Oxygen was actually toxic to the first life forms, and as levels of oxygen in the atmosphere increased, they had fulfilled one function; many of the then widely distributed first life forms died off in what was the first almost total mass extinction. They weren't totally eradicated, however, and we have recently discovered ancestors of those life forms. They are now known as extremophiles, found at great depth in the oceanic regions called black smokers. These are geothermal vents, areas of extreme heat where underwater volcanic vents pour gases containing suspended minerals. So some prokaryotes remained pretty much in their original form at the sea bed.

Whilst some prokaryotes, the extremophiles, remained relatively unchanged some went on to develop into different forms from the original. It is now, and only now, once life in the form of the first prokaryote has come into existence, that we can start to think about a process of development in life that has Darwinian evolution as a component of it. Modern bacteria are descendants of those original prokaryotes.

Bacteria retain a similar body form to their original prokaryotic ancestors.

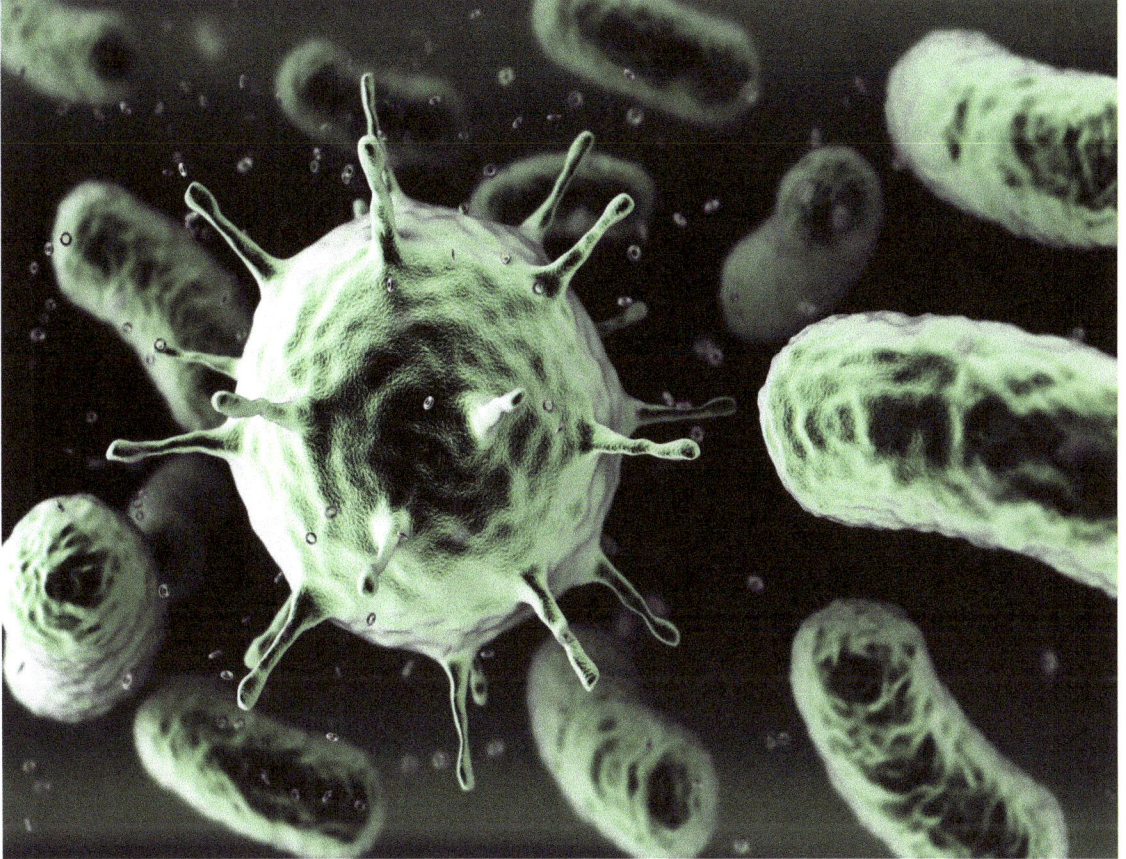

They are an example of a consequence of true Darwinian Evolution, retaining their original structure but having developed into different formats.

Eukaryotes Complex Single Cells

There is no apparent reason why a more complex cell should have spontaneously developed from the prokaryote. This more complex cell, known as a Eukaryote is, a completely different life form from our simple prokaryote.

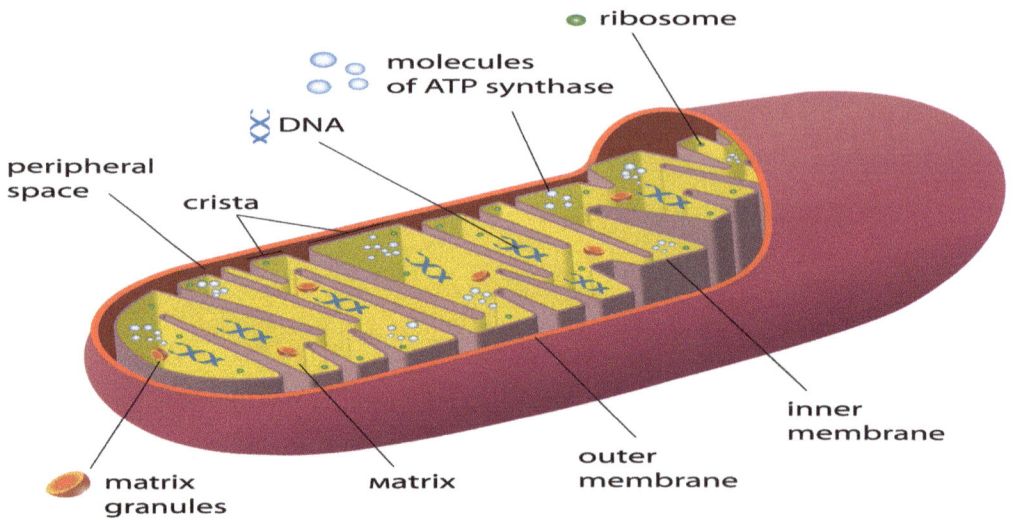

The body structure of a Eukaryote is a completely different from and far more complex than that of a Prokaryote.

The conventional argument for the occurrence of this radically developed complex single cell is that it occurred as a consequence of competition and Natural Selection.

As you can see from the diagrams of the two cells, the eukaryote and the prokaryote are entirely different lifeforms. I am not going to take you on a biology course but each of the individual "components" in the eukaryotic cell is far more complex than any computer.

The eukaryote is a totally different type of organism from the prokaryote. The new eukaryotic type of cell had a range of complex organelles to enable it better to organise itself and function. Virtually all of the DNA, reproductive directions, and instructions for the cells maintenance – everything that controlled functionality – had become housed in the cell's nucleus. The nucleus had of course itself first to develop in order to contain the DNA. The nucleus had then to move the DNA, which was responsible for its development into itself.

How does science account for the occurrence of a eukaryote? A much proposed suggestion is symbiotic engulfment. A prokaryotic cell engulfed another prokaryotic cell. The first ever form of cannibalistic murder. Or, to take a lighter note, two prokaryotes engaged in a dialogue appropriate to individual cells and agreed to work together in a symbiotic relationship. One engulfed the other, which became the nucleus.

The "other" cell became the more general cell. Together they cooperated on the development, worked on the additional components and on the design and development of further complexity. Organelles, sacs, became present to contain the various components additional to the nucleus such as mitochondria and chloroplasts. They formed the first ever complex cell: the eukaryote.

Yes, I know, this is what you expect to see in a science fiction novel, and no this isn't a science fiction concept. It is a serious science concept. An additional factor is that, by sheer coincidence, the newly developed cell was ideally and precisely designed to be suitable to form the basis for more complex multi-cellular life in the future. The eukaryote also continued the ongoing preparation of the atmosphere to be suitable for human habitation.

There is a slight problem with the idea, however. There is no evidence of any sort of a staged transition; different versions of these composite cells at various stages of transition is just one example. Interestingly, or perhaps more appropriately, amazingly, yet again we have a unique occurrence. Excuse me if the following words look familiar, by reference to an earlier paragraph about development. They are!

Prototypes must have come and gone; various versions came into existence until the self-assembled process was perfected. Where then are all the counterparts to the one prokaryote? A complex developmental process must inevitably have produced counterparts in the development process. They would have been less or differently developed functional versions and equally developed but slightly different versions. Identical versions with slightly differently developed genetic code and so on. Where are these counterparts? If there was no deliberate Intelligent Creation program that specifically created a unique unit how could there possibly have been just one cell?

However if there was an underlying purpose for life, an Intelligent Design driven by an Intelligent Creator seeking to progress onwards to human form then the scenario is different. The loosely distributed DNA of a prokaryote and its (relatively) basic format just didn't fit the bill. Something different was needed. Something that was differently organised and equipped with features that would enable progression to more complex multi-cellular life.

That something different was a eukaryote. Again a single cell organism, but one with its DNA, in a specifically defined format, concentrated in a central nucleus. The eukaryote also contained completely new features and concepts and was significantly more complex. Intelligent Design and Intelligent Creation suggests that as with the original one-off creation of life, the development of the more complex eukaryote was a specific part of the overall design plan to provide for humankind.

Interlude (2)

The absence of some sort of logical process, later peripheral evidence even, was a feature of the original occurrence of life, the prokaryote. We now see a similar occurrence with the eukaryote. Are we seeing a progressive series of events by an Intelligent Designer? The Earth is first created in a format that is suitable for life. It then, through a series of processes becomes adapted to be suitable for human life.

A simple prokaryote was specifically brought into existence. It had a role to play in helping to prepare the atmosphere for modern lifeforms including humans.

Was the prokaryote intended to serve two purposes however? A life form intended to start the clearance of the atmosphere and go on to develop under Natural Selection. There are numbers of instances in the developmental process of life where the potential occurrence of prototypes is seen. Was this a prototype development in anticipation of a need for more complex Darwinian evolutionary occurrences as life became increasingly complex?

After the prokaryotes we see the eukaryotes, completely different life forms from prokaryotes. There is neither tangible theory for eukaryotes having come into existence (unless you opt for symbiotic engulfment), nor any apparent reason why it should have occurred. Furthermore there is no peripheral evidence for its development were that to have occurred through some form of Darwinian evolution. Is this the specific introduction of Earth's second life form?

6. Multi-cellular Life

Soft Body Multi-cellular Life

The next stage of life's development was marked by the occurrence of life forms that comprised more than one individual cell: multi-cellular life forms. Cells that join to one another to become different specific lifeforms. Sponges for example, whilst an example of the most simple multi cellular creatures, are still significantly complex. Although sessile (fixed in one place) they have formed into a vastly complex life system. This includes entry and exit passages for nutrients, filtering mechanisms and sensors that enable them to differentiate substances in the water. They have internal cavities and specialised internal and external "wall" cells.

Sponges have gender, or indeed genders, as they can be simultaneously male and female and can play either role in sexual reproduction. They can also reproduce asexually by budding.

Sponges are an example of the very first form of multi cellular life.

Their bodies are not symmetrical as, say, a fish, with what is known as bilateral symmetry whereby they have clearly defined left and right sides. Sponges also do not have any defined internal organs. These additional and radical developmental complexities were to form in other later organisms. Like prokaryotic cells, however, sponges too, once extant, went on to develop many different species through Darwinian Evolution's concept of natural selection.

Interlude (3)

This is a totally radical change in the process of the development of life. The first two life forms considered, prokaryotes and eukaryotes, were unique individual life forms in themselves. Complex as they are and with individual DNA programs and body plans, they were nowhere near as complex as sponges. In fact our "simple" sponge is in a totally different playing field.

Sponges comprise collections of millions of eukaryotic cells which have somehow been organised to come together into yet another extremely complex new further life form. A massively complex symbiotic relationship between millions of individual cells, those individual cells themselves made up of equally countless numbers of individual components.

The individual cells are driven by their individual DNA and that composite collection of DNA somehow has come together to form a comprehensive body plan of the sponge, with which the individual cells cooperate. This thus is a second tier DNA. The cells are driven by their own DNA, and vast numbers of those individual cells come together under the direction of the sponge's DNA.

In terms of reproduction the sponge is a veritable hotbed of "sexual" activity. Individual cells are reproducing themselves copiously. Simultaneously, many of these are working towards the production of sperm for the sponge. Meanwhile the sponge itself is vexed with gender issues and whether to go for sexual or asexual reproduction. What a dilemma.

And this is a consequence of random selection of atomic elements?

Hard Body Parts and Organs

Timescales have been pretty much avoided until now because of the absence of reliable data. Whilst there is some evidential material in the fossil record of the life forms considered until now this has been limited. With the advent of hard body parts the fossil record, and other evidence has enabled timing of events with a greater degree of certainty.

The Cambrian Period 540-490 million years ago saw the occurrence of many significantly different life forms. The Burgess Shale is an area in the Canadian Rocky Mountains where a large number of fossils of different kinds have been found. Many of the fossils represent soft bodied creatures, but there were also many hard bodies and hard body parts also. For the first time heads, eyes and mouths appeared.

The developments during the Cambrian Period represent yet another totally unique development. Many of the creatures found at the Burgess Shale are unlike life forms we know of. They seem to represent species that occurred and then went extinct in a flash. Could these be more prototyping examples?

The creation and abandonment of prototypes that when tested were unsuitable for future intentions and scrapped.

We have previously considered prototyping and here we have yet another situation which would seem to be completely in line with the concept of Intelligent Design and subsequent Intelligent Creation. It is again difficult in the extreme to see how the occurrences at that time could have been the consequence of any form of Darwinian evolution. Leaving aside other considerations the short timescales involved themselves make Darwin Evolution improbable.

Further how could random processes clothe creatures like sponges with hard body parts and fit them with complexly designed limbs and eyes controlled by a complex organ such as the brain and its associated neurological and other control systems.

The occurrence of Burgess Shale creatures demonstrated many differential and new features of which just two are considered (again briefly) below. They have moved to DNA body plans that provided new innovations, the design and implementation of components and complex control mechanisms.

Organs

Animals don't just "have" organs. Organs are organised into a complex (that "complex" concept again) series of systems that interact together. A liver doesn't just come into existence in isolation as a liver. A similar consideration applies to other organs such as lungs and kidneys. Organs don't function in isolation but must be connected by systems, blood vessels and nerves. Organs also require nutrition in the form of blood.

Blood needs to be pumped. The organs need a complex body, far more complex than that of a sponge, to support them. Conversely, a complex body needs organs to physically support and sustain it.

Hard Body Parts

Hard body parts are not just a sequential development following through Natural Selection. They require the presence and thus the identification of the source of and the introduction of certain minerals in animals' bodies. This clearly leads to a need for the development of a specific biochemical mechanism by which creatres could extract minerals and other materials in appropriate proportions from their surroundings. Those minerals needed to be utilised appropriately in the host animals' bodies.

Hard body parts require the specific identification and utilisation of certain minerals.

The hard body parts developed at this time were not just basic skeletal and shell structures. Potentially aggressive structures such as teeth and claws also came into existence, along with correspondingly defensive ones such as spikes and plates.

Interlude (4)

If we accept that the occurrence of the sponge and what went before was a consequence of some form of direct Intelligent Intervention, then the latest developments simply represent a continuation of the process.

Life had become abundant with many different lifeforms and species. The structural basics were in place. If the underlying concept of an Intelligent Creator was to create a situation where human life could develop through a planned process of which natural selection was a part, then that point had been reached.

7. Life Moves to Land

To compare, for example, the challenge of the movement of life from water to land to the challenges of mans' ventures into space would present a significant difference of scale. Whilst the concepts are similar, the movement of life to land was uniquely far more challenging. For man it was necessary to manufacture complex space suits to retain the Earth environment necessary to survive in space. Life's movement to land required a similar need to retain an aqueous environment and an ability for respiration. The movement of life to land started from a very different development level and posed significantly different and greater challenge.

Life moved to land soon after the Cambrian period in the form of plants. The first plant life was in the form of algae and "relatively simple" plants. The description of plants as "simple" given their various components is clearly a massive understatement but serves its purpose.

The development of land life was not just the movement from the water to the land of the individual life forms themselves. It involved also a radical innovation in the social and ecological structure of all life forms that were to undertake the transitions. It required the creation of complex social and physical interactions between different types of life forms.

It is interesting to pause to consider the challenges placed on life in order for it to extract itself from an aqueous environment. Life as we know it cannot exist without some form of access to water. In the first instance, therefore, it was necessary for the emergent life forms to proactively design and then create a means of transporting an aqueous environment onto the land.

Equipment then needed to be created and manufactured that enabled the transition from respiration in water to respiration in an externally water-free environment. Additionally, it would have been necessary to instigate a program of physical development to facilitate dealing with the change of pressure in an aqueous environment to that of land.

Having devised a mechanism to get out of the water, what kind of lunch box supplies did our super fit emergent ancestors bring with them? There were no cafes or restaurants, so what did they eat? Finally, our super fit emerging explorer encased in her water cocoon would have had to figure out how to have sex and produce offspring that were similarly endowed.

Some of the first land animals were ancestors to modern crabs. They had developed that water carrying environment and other necessary features which enabled respiration and functionality on land. Many had strongly developed shell structures.

There is much debate about how the predecessors to this modern day crab could possibly have developed the mechanism necessary for life above and below water.

Insects also arrived early on the scene as specific invertebrate animals with jointed legs.

Insects can be tiny yet they are as structurally complete as larger animals.

Take a look at the insect picture, so small a creature yet so very complex.

They had very functional complex eyes and were the first animals to develop powered, yes powered flight. We humans first managed to create powered flying machines just over 100 years ago, and even today the flight of insects and other winged creatures can be technically more complex than that of humans. Some of the creatures that fly have complex multiple sets of wings and can perform aerial acrobatics we can but dream of.

The transition of animals that moved across land as opposed

to swimming in water marked another innovation of a specific type. Fish move through water using fins and are in part supported by the water. The many types and varieties of animals that populated dry land faced the challenge of gravity whilst unsupported by water. This required the design and development of yet new features including complex leg structures and muscular structures that could cope with this change in the environment. Ongoing it was necessary to develop a mechanism for the production of eggs contained within the body a feature requiring an internal to the body birth development process.

The changes at this time, and indeed throughout the entire development process of life were not restricted just to the need for individual organisms to develop themselves as specific species and individual entities. It involved a far wider and further reaching set of requirements. A whole new regime of animal and ecological interaction was needed. This required the complex cooperation of all the many, many aspects of the planet Earth. Seasons, climate, distribution of nutrients by river flow to mention just three. It is a widely believed that disruptions in just one species can have far reaching consequences for other species.

The "butterfly effect" for example is a complex aspect of the mathematical theory of chaos. It holds that very small changes in one situation can result in complex subsequent changes on a far wider scale. It talks about the concept that a butterfly flapping its wings can cause knock on effects that escalate and lead to the occurrence of a hurricane somewhere far away.

The Butterfuly Effect suggests that a single small change can result in a large effect over time.

Thinking about this further concept of the butterfly effect it is worth bearing in mind our overall concept of "natural" events. In relation to the concept of Intelligent Design it is not necessary to be restricted to thoughts of individual phenomena, the occurrence of specific new lifeforms. In the concept of Intelligent Design any event can be attributed thereto, including the ecology of the planet and animal interactions therewith.

The more deeply we think about these things the more complex and intimate the interactions are seen to be.

Moving on to a further development dinosaurs first appeared on Earth some 250 million years ago and came to dominate much of the planet. Some were small, some massively large. Some were docile herbivores others aggressive carnivores. Interestingly some of the dinosaurs had features similar to those of humans. These features included walking in a semi upright bipedal way. Their forelimbs were not weight bearing and did not needed to facilitate walking and running. There is no apparent reason why those forelimbs could not have developed into structures similar to human arms and hands.

We have no way of knowing the challenges dinosaurs could have posed to the ongoing development of mammals and hence humans. It seems a reasonable proposition that such large creatures, many of which were well armoured and protected would have presented a significant challenge to the generally smaller softer bodied mammals from which human kind developed physically.

There must inevitably have been competition amongst the two life forms. It clearly cannot be possible to know if, had that competition been allowed to continue, humans could have developed as they did. The competition between the two life forms did not however continue and a mass extinction of the dinosaurs occurred leaving the way clear for the development of mammals and consequently humans It is believed that the mass extinction occurred as the result of an impact of the Earth with a relatively large asteroid or comet.

Thinking about impacts with the Earth you will see in Annex 1 how the Earth's Moon formed as a consequence of a similar impact event between the Earth and another spatial body, albeit the impact was of a different magnitude. This impact event occurred in the early stages of Earth's development when conditions that would enable the Earth to be suitable for the development of life itself were occurring.

We have seen that flight developed in insects at the early stages of the occupation of the land by animals. Insect flight is a particular style of flight that is suitable for these small creatures. The basic technology underlying all winged flight is the same. The wing is formed so as to create a lower air pressure above it than below. It is clearly a technical design in itself. Whilst the principle is the same for all winged creatures, flight in larger creatures such as winged dinosaurs is an entirely different technological concept from flight in insects.

The pterodactyl was a small, now extinct, flying reptile.

Bird flight is an interesting example of specific design features necessary for functionality. Backwards and forwards pointing claws were necessary and needed to be designed to be able to perch on branches. It's not a lot of good being able to get up there if you can't stay up there. Other essential complex features, mechanical and technical ones, related to forelimb (wing) relationships with other body parts and bone and skeletal structure. These were all innovative designs. Another noticeable feature is the development of wing and tail feathers with specific individual flight functionality.

In getting to this point in the book, we have looked at many concepts of life. Might this be a good time to look at an example of what the beauty of what life actually is?

The humming bird for example is a magnificent example of beauty and complexity.

The relationship between flowers and the humming bird is an excellent example of mutual reciprocity.

It can hover, fly at 15mph and stop instantaneously, fly backwards and upside down. Take a look at the humming bird about to interact with a flower Look at its beautiful yet complex design. It and the flowers with which it interacts with are a perfect example of a beautifully designed system. They function in an absolute perfection of harmony and symmetry.

Humans belong to the animal group of mammals and started to develop alongside dinosaurs some 250 million years ago. Mammals are characterised by features that can be differentiated from other species. They have hair and feed their young with mammary gland milk. They are part of the ongoing development of yet different life forms that were to lead to humans.

Emergent from the plethora of different life forms that came into existence was a very distinct order called primates. These included apes and monkeys. It is in relation to this order that we can start thinking distinctly about the development of humans.

We are not going to consider any form of detail the nature of developments up to and including the emergence of the primates. There are so very many differential factors, complexities and design issues that have occurred that it is simply unfeasible to do so.

8. Animal and Plant Relationships

Ongoing this book works towards the specific development of human life and what it is that makes humans different from other animals. We are shortly going to focus on the development of human life, but before doing so, it would be neglectful not to pay some attention to the interaction between animals and plants. After all, without plants there would be no animals and thus no human beings.

We don't just "have" plants and animals. Plants have been so beautifully and specifically developed to work in absolute congruence with animals, yet another carefully designed symbiotic relationship. Plants were absolutely essential to the original development of animal life and remain equally as essential to the continued existence of animal life and we humans.

Plants depend on a supply of energy to enable their life processes. All energy, and it's worth a repetition, "all energy" delivered to the Earth and used in the manufacture of food consumed by animals comes from the electromagnetic radiation delivered from space. The source of this energy is the sun.

Electromagnetic radiation for practical purposes here is, in essence, light, if you like, of the colours of the rainbow. Certain colours are absorbed by plants and others cannot be absorbed and are reflected. Many plants are seen to be green for example because they cannot absorb green light which is reflected.

Photosynthesis is the absorption of the energy contained in

the sunlight by plants which is used to drive chemical processes that enable them to grow.

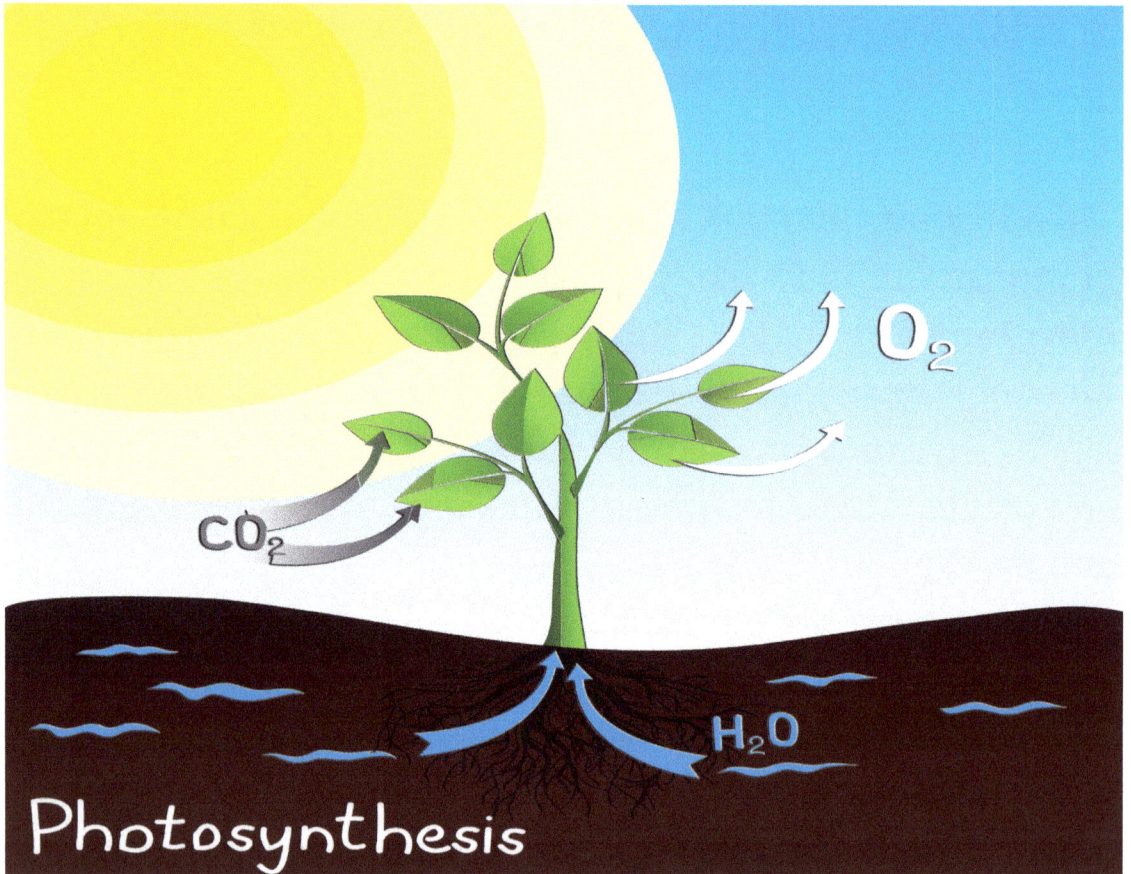

Animals are unable to draw energy from the sun and depend on plants for their nutrition.

Animals are unable to undertake photosynthesis and cannot therefore use the energy from the sun.

Without the ability to undertake photosynthesis animals must ultimately depend entirely on plants for nutrition. Plants are chemical factories with a specifically designed function. They draw other nutrients and water from the soil, grow, and are available to be eaten by animals. Some animals, herbivores consume only plants. Other animals, omnivores, consume both plants and animals. Carnivores such as lions and tigers eat only other animals, both herbivores and omnivores.

Theoretically a giraffe is a herbivore. Like much in life it isn't that simple. When munching leaves it consumes animals, insects, making it technically an omnivore.

Whatever an animal eats therefore, it has, either directly or indirectly, consumed food that had its origins in plant life. No animal existing today could be alive were it not for the existence of plants. All animal life (and human life is included in this) is totally dependent on plants for its existence.

The relationship between animals and plants is at least in part reciprocal. The supply of nutrients available to plants in a particular area of soil, indeed every area of soil, is finite. If those nutrients are not replaced then they will not be available to support plant life at that location and no further plants will grow successfully there.

Plants are generally sessile. However even were they not

sessile and could move about the land freely, they would eventually use up those nutrients in the soil everywhere if those nutrients were not replaced. Those nutrients thus need to be replenished and this is where the animal / plant cycle cuts in. Animal droppings and decayed animal remains, along with those from decayed plant growth, make a contribution to nutrition available in the soil.

It is a mutual process and animals also play a significant direct role in plant reproduction. There are many examples. Herbivores consume plants, some of which contain seeds which are indigestible by the animal. Animals move about the land, depositing those seeds in different locations where they are able to start fresh growth. Plants are pollinated by animals (pollen is moved from the male to the female) enabling fertilisation and reproduction.

The planetary ecosystem is the way in which plant and animal life on Earth interact with each other and their environment so that a system of balance is maintained. The interactions between animals and plants are however just part of the complex relationship between the two. Complex as the sum of the animal and plant interactions are they are just a small part of a vastly complex planetary ecosystem. We touched briefly on this earlier when we mentioned the "butterfly effect".

The design of the ecosystem is such that the balance of life retained on the planet has created the appropriate environment for the emergence of physical human life, and to sustain it to the modern day. The reference to "physical" human life is deliberate and will be explained presently.

The interactions between the varying components of the Earth are vastly complex. An interesting example is found in the Sahara Desert sand, which contains significant volumes of microorganisms and fossilised larger animal remains broken down into sand. Large volumes of this enriched sand are picked up by the wind and carried to the Amazon rainforest where they are discharged in the rain as nutrition for the soil. There are many other means of nutrient distribution. This distribution of nutrients is a significant factor in the ongoing existence of the rainforests.

Another example of such large scale planetary interaction is the distribution of sediment by the outflowing River Nile. On a different scale the ocean and their currents play a major role in helping control the Earth's temperature and general climate. The southern part of the World, the Southern Hemisphere is generally warmer than the North. This is because the south contains more water and less land than the north. Water heats up more quickly and cools more slowly than land.

9. A Definition of Human Life

Believe it or not, the earth is currently in the grip of an Ice Age, and the relative warmth and stability we are presently enjoying is a pause in the ice age known as an interglacial period. This current interglacial period, which started some 12,000 years ago, saw the radical occurrence of a new type of human being that was different from all that had gone before, "Secondary Evolution Human Beings" (SEHBs). If what has gone before has stretched your comfort zone, then be prepared to have it stretched even further.

It is necessary to be absolutely clear here that we are discussing a pivotal concept. This is a completely unique human animal with an occurrence defined in a completely different way from any animal that had existed before. Until now we have considered purely physical occurrences. When we come to consider the SEHB will be considering whether events leading to we human beings coming uniquely into existence, can possibly be attributed to physical occurrences. You will see that it is difficult in the extreme to conclude that they could. If not physical then what, the Big Question!

The physical development of humans can, in the context of a physical format, be traced through the development of some twenty different types of humans over the past several million years. The modern physical species of humans came into existence about 200,000 years ago in the form of *Homo sapiens*.

The generally portrayed picture of human development is one of a historically well-developed species of humans. In this picture human development, from the distant past to the present time, has been a linear one. Humans are considered to have been already significantly differently developed from other animals at the start of the interglacial period. That they lived in small communities and had developed rudimentary skills is interpreted that they were primed ready to launch into modern man via Darwin Evolution.

Is this correct or even possible?

Neither the question nor the answer are straightforward. At the start of the interglacial period something occurred that had the effect of creating a completely different type of of life form, the human being. The occurrence was different from that which had happened with the unique inception of other life forms, be they cells, sponges or more complex forms. The occurrence represented the unique appearance of a life form that was distinguished by something other than its physical form. The physical human life form remained physically the same. What changed was the underlying nature of the life form. It had something different, something extra.

Before we go on to discuss the changes, let us first consider the previous thought that humans were not significantly differently developed from other animals at the start of the interglacial period. There are numbers of characteristics that, ostensibly, it can be claimed differentiate humans from other animals.

A number of the principal differentiating factors often quoted are listed below. Perhaps you can think of others. It is, however, the case that with a ccouple of exceptions, all early human behaviour in the period before the interglacial period and listed below can be seen in at least one other type of animal, and frequently many species.

Erect structure and bipedal motion.

These are general characteristic shared in varying degrees by dinosaurs, gorillas and chimpanzees. The Bonobo chimpanzee living in the forests of the Democratic Republic of Congo spends much of its time fully upright.

A Bonobo mother with child.

Tool use.

Tool use is widespread and readily catalogued in the animal kingdom, from gorillas to wasps. Wasps dig burrows in which to lay eggs, place dead insects they have killed in the burrow, fill the burrow in and then hold small twigs in their mandibles to close the ground down.

Skilled Hunting.

Assassin bugs use the skill of baiting. They bait prey by placing dead termite carcasses in places where other live termites are vulnerable. They then capture and consume those termites when they try to eat the carcass themselves. There are many, many skilled hunting animals, even the odd carnivorous plant; the Venus Fly Trap, for example.

Communal Living.

Just think of ants' nests with their infant nurseries, bee hives and lions. Then all the other communities in the animal kingdom. Taking termites just as an example, they are organised into specifically defined roles such as workers and soldiers. Offspring are collectively cared for and nurtured. They communicate extensively by way of a language of signals, in colonies of up to a million or more.

Many, many animal species live in communities. There would be nothing illogical for early humans to have been living in groups at relatively fixed locations also.

Courtship.

Courtship is widespread amongst animals. Male Empididae flies wrap the carcasses of organisms they kill in silk and offer them to females to encourage sex. Females in turn have developed various attractive ornaments, such as changes to wing colour, to encourage males. Not ornaments, perhaps, that we might consider attractive. The crane employs the most complex dance in the whole animal kingdom, one that is still not understood by man. Ferocious lions enjoy courtship too.

Lions have their softer side.

Infidelity

Gibbons mate for life. They can also be unfaithful, as can other non-human animals that ostensibly mate for life. Infidelity was clearly not a unique human trait either.

Human Migration out of Africa.

Migration is a very common feature of the animal kingdom. Lemmings migrate. That they commit suicide en masse, however, is a misconception. When population groups become too dense, their migrations may unintentionally end in disaster.

Other Unique Characteristics

The list could go on, but objectively it is difficult in the extreme to see anything tangible that sets humans significantly apart from other animals in the period leading up to the last interglacial period some 12,000 years ago.

Moving on again let us think about a colony of animals that existed at the start of the current interglacial period and displayed the following characteristics.

1. Lived in mutually reciprocal communal colonies of 50 or more.
2. Willingly shared food with one another.
3. Stripped plant stems and used them as tools to fish with.
4. Invaded the territories of others and staged deliberate war.
5. Communicated verbally.

6. Were sexually prolific.

7. Et al.

Was this a primitive human colony? Indeed it was, but it is also a colony of chimpanzees then, and today. Humans moved on, chimps didn't.

There were indeed skills that were unique to early humans, cave painting and the use of fire. These are skills that are unique to humans. Is painting more skilful however that a spider spinning a web or a bird building a nest? A particular animal type having a unique skill is not uncommon at all in the animal kingdom.

10. The Occurence of SEHBs

Human beings became defined in their present form in the blink of an eye in relation to archaeological timescales. Yet it was in that instant, a period of a totally different Secondary Evolution, that saw the coming into existence of nature of human life as we now know it.

Before trying to define SEHBs let us look at the basic animal senses, of which there are five traditional ones. These are: sight, touch, sound, smell and taste. All animals including humans depend on one or more of these for reproduction and survival. Reproduction and survival are the two basic driving forces underlying all animal behaviour.

The senses an individual animal uses are those that enable it physically to locate and consume nutrients and reproduce. It's a bit of a basic definition but think of an animal in a field, a rabbit maybe. What does it do? Well, you probably have heard of the concept breeding like a rabbit; hopefully that's self-explanatory. What else does it do in addition to breeding? It eats and survives.

Exactly the same concepts apply to animals we may regard as more complex: an eagle, a lion, and in particular the previously mentioned chimpanzee. They are all, including chimpanzee colonies, driven by the same driving force, to eat, survive and reproduce. SEHBs are no exception, from a specifically physical perspective.

Being absolutely blunt and basic, no matter how attractive or otherwise animals generally may be, they are, like plants, chemical factories. They function differently from plants but chemical factories they (including us) are. As we considered a while ago there are ninety naturally occurring chemical elements, many of which form the basis of an animal body and its nutritional needs. The chemical factory that is an animal consumes nutrition, extracts the chemical elements that it needs to physically function, and ejects the residue as waste product. It comprises a sequence of automated processes.

All of the animal's physical functions within its body are driven by chemistry. Vision for example is the chemical process of converting the light that enters an animal's eye into messages that the brain can understand. The conversion of messages in the brain is a chemical process. The decision of an animal to move in a particular direction is driven by complex chemical processes associated either with a physical need such as to obtain nutrition. The process extends beyond an animal's immediate need. It extends to animals' need to interact, say, with its counterparts. Wolves in a pack hunting prey, for example. The interaction between one wolf and its colleagues; its anticipation of what another wolf will do.

Wolves hunting in packs need to be able to communicate in different ways.

All are driven by the demands of the chemical factory. It maybe horribly blunt but it's true and it applies to every animal and every basic animal instinct.

As noted, the foregoing applies to SEHBs equally. SEHBs however are different in that they have something more, indeed much more. Something happened at the time of the Secondary Evolution that caused SEHBs to become very different from their previous counterparts. Life for SEHBs became not limited just to the need just for survival and reproduction, something else developed instantaneously.

SEHBs became human with all the complexities that define human life forms. How can the complexities of humanity be defined in words? SEHBs have consciousness for example, a concept which, like other concepts associated with it, is impossible adequately to explain. Consciousness generally may be defined as being aware of one's surroundings. Awareness can be explained as having perception of a situation. And perception is defined as being aware. And so it goes on. They are just words, abstract concepts. The use of abstract concepts does not enable us to quantify what we seek to represent here.

Moving on and skipping the abstract let us think about aspects that define humans. Think about the following example.

* Walking along the street an individual, an observer, finds an injured person on the edge of the road, the subject of a hit and run incident.

* The observer experiences a feeling of concern about a fellow person.

* The observer kneels alongside the injured person to see if it is possible to assist the person.

* The observer goes further and puts their jacket over the injured person, getting it stained with blood and ruined.

* The observer comforts the person, who is dying, and is given a message by the dying person to give to a relative.

* The observer later locates the relative and gives them the message, briefly sharing the relative's grief. The observer may even attend the funeral sitting in the background.

The actions and feelings of the hypothetical observer were totally altruistic, showing a selfless concern for the third party. So what occurred in that one relatively short period of time when the two people interacted? In respect of an isolated incident, the observer was involved in numerous occurrences, at least five complex situations which, in part at least, defined her as a human being. The observer's emotions were unique to humans.

There could be many other emotions associated with the same incident. If the deceased person's injury was the result of a hit and run accident, the observer might well feel anger at the driver. If it had been possible to make a note of the departing vehicle's registration number the observer may well have done so and passed it on to the police so that justice could be done. We then discover the further concept of justice, and a quest therefore.

The montage of emotions and feelings experienced in the previous example are just an example of one set of many emotions that humans can experience.

The emotions suggested in the previously considered hypothetical example were indeed poignant. The feelings evoked may however, pass relatively quickly with a speed depending on the individual person. We can reflect that even the fact that the emotions will pass at variable speeds depending on the individual is a further aspect of the definition of humans in terms of individuality.

Had, however, the subject of the hit and run incident been a non human animal, then whilst the observer might have experienced similar emotions, they might be of different nature, manner and magnitude.

Had the deceased person been close to the observer, a relative perhaps, the emotions might, or might not, be far deeper seated and lasting. Whether, and how much so, would be dependent on the nature and strength of the relationship.

It does not require any effort to go into further consideration about the incalculable number of variations of emotional situations that can occur.

Relationships themselves can be so very different. The relationships between parents and children, uncles and aunts for example. The variations and permutations are incalculable. But emotions, poignant and defining of humanity as they are, are just one tiny aspect of what makes us human.

Let us move on from emotions to another aspect that defines humans: communication. Communication can be defined as a means of sending and receiving data, something that all animals do. At the event of the Secondary Evolution, SEHBs developed a concept of communication that was not restricted to the need to satisfy the basic driving force of survival and reproduction.

Human communication involves a specific mechanism that can be, in relation to English for example, portrayed by punctuation and other symbols and the twenty six letters of the alphabet that make up the English language. The English language itself comprises more than a million words with more than ten added each day. Individual words can in isolation have more than one meaning and are also grouped interchangeably into different categories, such as nouns verbs and adjectives. Nouns for example can be further sorted into sub groups, common nouns and proper nouns, the same word written differently depending on context.

Much is made of the fact that *Homo sapiens* had developed larger brains over time in the period leading to the Secondary Evolution. Think about this in the context of Intelligent Design, however. Remember that unlike the occurrence of our promiscuous sponge, the occurrence of the SEHB did not represent a physical event. The SEHB was physically identical to the human that existed immediately before. The change was in the non-physical nature of the SEHB.

We have already considered the concept of prototyping and preparation. If logically an Intelligent Creator was planning to provide for a SEHB, the preparation of the physical nature of the SEHBs to be able to cope with the new demands placed on them would be completely logical. These would have included the development of necessary cranial capacity and other assets such as the dexterity necessary to exist on foot without physically developed means of defence.

In a similar vein, the creation of the twenty or so prototypes of modern humans would have been a logical process as in developments leading to *Homo sapiens* themselves.

Getting back to communication, language is not restricted to the specific symbols that it contains. Two sets of symbols constructed in the same sequence can have different meanings, "captor grills captive" for example. Was the captive questioned intensely or cooked? You know the answer. So language does not just comprise statements, but can require an even deeper intuitive understanding of the various nuances of statements.

The need for a deeper interpretation is not, of course, restricted to the written word, but extends to nuances of the spoken one. Suppose you were discussing horse breeding with a breeder, who explained to you in respect of a particular incident, "The mating between the mare and the stallion failed sothe vet stepped in and sorted it." What did the vet step into? What did he sort? Did he physically take over from the stallion or arrange for things to proceed artificially? What would physically taking over mean? The complexities of communication go on, and on, and indeed on.

Body language combined with context with what is said, is a further factor. One person saying to another " I like you" can be projected and interpreted in many ways, depending on how and where it is said and the bodily actions associated with the words.

The toss of the head and the glint or otherwise in the eye being just two examples.

Hopefully this speaks for itself.

I belabour this one no further and leave it to your experience and interpretation. Communication can be in the form of body language requiring no spoken words at all.

As with every individual concept in this book, it is possible to write at great length about the very many other aspects that distinguish humans. Each is a subject in itself and it is left to you to ponder. Think just about music and mathematics, to start. Then there is abstract thought. Then even the fact that you are able to ponder. And so it goes on.

The occurrence of the SEHB at the start of the interglacial period was completely radical by reference to what had gone before. Historically occurrences had been physical.

What brought about the occurrence of the SEHB?

No one knows even where the nature of these defining factors that make humans human are located. Are they physical or non-physical? From where are they derived, from within the physical body, and if not where?

Medical science may argue for example that those defining factors are all physical, part of the structure and function of the brain and associated body parts. Take the emotions we just considered as an example of something that defines humans. That they occur as part of the physical entity it is claimed can be demonstrated, for example, by giving someone a drug which can be seen to change the nature of the emotions, it is argued. Following on from this the argument has it that emotions and changes therein must derive from the physicality of the body.

Now in terms of day to day human function we are designed in such a way that the occurrence of emotions and their expression is part of our functionality. They are involved in person to person interactions that define humans and must be physically expressed.

Emotions are expressed in many different ways.

That we physically express emotions does not however require that they be derived from within the physical body.

Is what we perceive to be a change of emotion after taking a drug just a change in the way a drug physically requires the body to express the emotion, rather than a change in the underlying emotion? What happens to the emotion after the drug wears off?

Suppose that emotions themselves are non-physical, originating from a non-physical element of a SEHB. Is the brain simply a vessel, a mechanism for enabling the physical expression of those non-physical emotions to other physical humans?

Everything that makes us human started at the time of the Secondary Evolution. It is difficult to imagine any way in which the non-physical changes involved becoming a SEHB can possibly be attributed to physical changes to, and developments of the physical body. If you go for the Darwin Evolution concept that is something that occurs over a lengthy period of time. The changes are just too dynamic and far reaching to be a consequence of this.

So what happened? How? Why? All the changes occurred instantaneously. There is a vast plethora of evidence that demonstrates the instantaneity of events at just 12,000 years ago.

Such evidence occurs, for example, in the occurrence of the oldest known temple, the Gobekli Tepe Temple in Turkey, built, 11,500 year ago with complex carvings it was constructed at the very start of the Secondary Evolution. There was nothing even similar before. The ability to undertake its design and construction just occurred.

Moving on from the immediate incident of the Secondary Evolution, in just an incredibly short period, the entire world of early SEHBs witnessed a colossal change.

1. Complex, often mathematically based structures were built. Temples on land and carved into rocks were created as were pyramids, and many, many different kinds of construction. All these were essentially geared to some form of relationship with an Intelligent Creator.

2. Civilisations occurred in many different forms and complexities, with many different languages and concepts. Again these civilisations, without exception allude to an underlying theme of relationships with an Intelligent Creator.

Interlude (5)

What are we to make of all of this? Could the myriad of developments that occurred in the blink of an eye possibly be proposed as the consequence of yet another random occurrence?

If that concept is rejected then it would seem that the only alternative is that this, development in the human story, along with everything that lead to it, is a direct result of Intelligent Design, Intervention processes at the behest of an Intelligent Creator.

Furthermore, where are the unique factors that define humans to be found? If they do not derive from the physical elements of the human body then where?

11. Conclusion

There are numbers of ways of try and define certainty. The one that can be used here is the statistical one. The likelihood whether the events discussed previously could have occurred as a consequence of a random series of events. As we discussed previously in relation to consciousness we can endlessly play on words to try and define likelihood and probability. Rather it is better to use some examples.

If we toss five coins in the air simultaneously they will varyingly land heads up and heads down. If we toss the five coins simultaneously a million times it is likely that all five will land heads up at the same time on a significant number of occasions. The more coins we toss a million times the less likely it is that they will all land heads up at the same time.

There are many permutations which could be applied to the concept of random occurrences in relation to what we are considering here. Conventional creation arguments have it that the number of random chemical interactions that might have occurred on the early earth over the long time periods involved make it inevitable that the First Life occurred. It is technically a feasible argument and it would be foolish to seek to deny this possibility in isolation.

Conventional arguments have it that subsequent occurrences of different life forms also fall into the "consequence" of statistics combined with concepts of Darwin Evolution. It is for you the reader to consider whether the possibility of such occurrences moves away from the possibility of a reasoned technical argument, to a concept that is so completely distant from that, as to be completely infeasible.

In considering this you may also wish to have regard to the various other factors that we have discussed. If you conclude that your presence here could not have been a consequence of the random events proposed then there seems to be only one alternative conclusion.

Life and the occurrence of SEHBs is a direct consequence Intelligent Creation, consequent upon an Intelligent Design process.

There is an Intelligent Creator!

12. The Future

We reflected early on in this book that the Solar System is "ours". A unique place physically isolated from the rest of the Universe by a (currently) unsurmountable physical distance. We speculate generally about there being other life in the Universe. Could it be that planets such as ours are indeed common? Might many such planets have been seeded with life by the Intelligent Creator? Might life on any such planets have the same opportunity to develop as us so that the concept of Intelligent Driven Design driven is not restricted to us hear on Erath?

An egg provides all that is necessary to support an embryo until birth, so that the newly hatched creature can move on to develop and realise its full potential. Were we humans and the planet we occupy at an embryonic stage with the event of SEHBs?

Did we have, do we still have an Embryonic Earth?

At the time of the Secondary Evolution the Earth was, from a physical perspective, a perfect and idyllic place for newly occurred SEHBs to develop. It provided an embryonic environment with all the resources necessary to support SEHBs through to "maturity".

There was and continues to be an absolute abundance of natural resources, which meant there was little if any need to resort to the use of fossil fuel for energy and power. The sun alone delivers hundreds of watts of solar energy per square metre ongoing.

There is no need for food shortages. There are, as just an example, vast areas of desert available for cultivation. This is not a fantasy. As long ago as 1913 the engineer Frank Shuman developed a project to irrigate areas around Cairo. It was deemed very promising but it was brought to a premature end by the outbreak of the First World War.

Let's get to it however! The Earth, our precious planet is being ravaged and destroyed. We are not talking about some distant descendants of ours! We are talking about us, our children and our grandchildren! Forget the grandchildren even and think about us and our children! Albert Einstein famously wrote, "I do not know what weapons World War III will be fought with, but World War IV will be fought with sticks and stones".

We must prevent the almost inevitable.

As resources become increasingly sparse, nations will fight for them.

If we carry on as we are, pushing to the Rubicon, the destruction of human (and other) life on the planet is not something that will occur at some time in the distant future! The destruction will be a holocaust far worse than anything ever experienced, horrific as the last holocaust was. It is imminent **NOW!** You and I could, if we survive the holocaust see our days out living on berries!

The solution absolutely does **NOT** lie in some nebulous

concept of government policies of changing something or other by some abstract amount over some equally abstract period of time!

Things must change!

NOW!

Quangos and the like are no use here (or indeed anywhere else) Every government in every "civilised" country needs to seek out individuals who are both passionate about change and are simultaneously able to work in cooperation to bring it about. Leads need to be taken.

Resources need to be developed and deployed to "less civilised", "less advanced" perhaps countries so that it is to their definable advantage to move forwards.

An impossible pipe dream, a fantasy perhaps. But if it doesn't happen we will be competing with other animals on unequal terms (to our disadvantage) for food!

Could we compete with this squirrel stripped of our modern technology?

ANNEX I

Earth's Features and its Location in Space

The Habitable Zone of the Earth

The Earth is located in a collection of planets orbiting the Sun in a planetary system known as the Solar System. The planets orbit the Sun in almost circular orbits from Mercury outwards, held in place by gravity, a force that we discuss later. The Earth is the third planet out from the Sun and occupies an orbit that is very clearly and specifically located in an area of space, falling within what is described as the Goldilocks Zone.

This Goldilocks Zone, the Habitable Zone in which the Earth orbits is not too hot and not too cold for life to exist. It is an orbit located within a specific range of distances from the sun that enables water to exist in all three forms: a solid as ice, a liquid as water and as a gas. The ability of water to exist in all three forms on the planet is, as you will see, essential to the existence of life as we know it. The range of distances from the sun in which the Habitable Zone exists is fairly narrowly defined; too close and all water would boil into gas, too far away and all water would be ice. The Earth orbits the sun at a radius of 150 million kilometres. The radius of the Sun to the edge of the Solar System can be measured in numbers of different ways. One estimate is that it is 150,000 million kilometres.

Albeit potentially very coincidental, we have our first piece of evidence to suggest the deliberate location of the Earth in a specific orbit around the Sun. The next few pieces of data that we are going to look at are going to be ones that are increasingly suggestive of the Intelligence of Creation.

Debris, Rocks and Gravity

In the previous example we simply saw the fortuitous location of the Earth's orbit in relation to its ability to sustain life. Now we move on to see just how it got there. We saw in the main text how the Solar System formed through a series of collisions between rocks and debris.

Some collisions were destructive, breaking the rocks apart; others constructive, sticking together and creating ever larger rocks. Eventually the planets as we know them formed in orbits around the Sun.

The position of the planets hasn't always been as it is now and it is believed that some used to be closer to the Sun and others further away. Some crossover of orbits occurred where planets actually changed position relative to the sun. It is a fact that the orbits of the outer planets as they are now provides a significant protective shield to the Earth from meteors and comets that might otherwise impact the Earth. Such bodies are caught by the gravity of those planets and impact on them rather than being allowed to continue on their trajectory and potentially impact with the Earth.

Colliding pool balls cause the other balls on the table to adopt certain locations. In the same way each collision between the early Earth and the other bodies in the Solar System were instrumental in defining the Earth's final orbit.

Gravity is a force of attraction between two or more bodies, for example the Sun and the Earth. It is a mutually attractive force drawing the bodies closer together. The sun and the planets mutually draw each other to and from one another as a result of the gravitational forces they exert on each other. The earth is held in its particular orbit in part by the balance between the attractive gravitational force of the sun "versus" the "opposing" gravitational forces of the other planets in their orbits. It is a very finely define location.

It was a specific result of the combination collisions and gravity that caused the Earth to be where it is in the Habitable Zone. Subject to some fine tuning, as we see next.

Earth's Moon

The moon formed as a result of a collision between the early, and what had become the relatively stable Earth, and another large body, possibly a planet.

The collision that caused the occurrence of the Moon was very important, perhaps even essential for human life on Earth.

The body struck the Earth a glancing blow. This threw off a large volume of debris which eventually formed the Moon.

The collision did not, however, just form the Moon, but was a further factor in defining the Earth's position firmly in the Habitable Zone

The immediate result of the impact saw the Moon's position much closer to the Earth than it is now. Gravitational forces have resulted in the Moon's migration from the Earth at a current rate of four centimetres per annum. The current beneficial position of the Moon is a direct result of its initial positioning following its original formation, and the rate of its migration from the Earth.

The Moon's current position is important and beneficial to the Earth. It plays an important role in maintaining and stabilising the Earth's rotational axis. It also causes the tides as we know them. These tides are instrumental in moving heat from the Equator, helping to stabilise the Earth's temperature. Tides also distribute material about the Earth.

At almost 25% the diameter of the Earth, the size relationship between Earth and the Moon is unique by comparison with all other known planets. Had the Moon been a different size, then its previously described relationship with the Earth would have been very different.

Earth's Atmosphere

The Troposphere, that part of the Earth's atmosphere in which we live and breathe, comprises some 21% oxygen and you will almost certainly be aware how important oxygen is to life. Earth is the only known planet that has an atmosphere capable of sustaining us. As you will have seen the atmosphere of early Earth did not contain oxygen. The occurrence of oxygen on Earth occurred over a long period of time, as the planet started to become prepared for our existence.

Earth's 23.5 Degree Tilt

The Earth's orbit is elliptical (not quite round), so the distance from the Earth to the Sun varies slightly over the year. It is common misconception that this distance of some 5% is responsible for the seasons enjoyed on Earth. Whilst this has a small effect, the seasons we enjoy are mainly the result of yet another specific phenomenon associated with the creation of Earth.

The Earth's rotational axis, the angle of tilt relative to its axis of orbit around the Sun, is fixed at an angle of approximately 23.5 degrees relative to the Sun.

Whilst the Earth's tilt was probably not essential to human life it has brought many important consequences.

The technical explanation for this is not straightforward and is well explained in many other publications. Essentially, when the Earth is on one side of the Sun, the north of the Earth is closest to Sun. Some six months later when the Earth is on the opposite side of the Sun, the south of the Earth is the closest.

So at what is called Summer Solstice, June 22, the north of the globe is most pointed to the Sun so it receives a maximum heat from the sun (subject of course to other climatic conditions). As the Earth continues in its orbit, on 22 December the opposite occurs.

The tilt as defined is a factor of importance in the nature of life as it is, although in the interests of objectivity it is not possible to claim in any way that life's existence was or is dependent on it.

Earth's Protective Force Field

The Sun by virtue of its nature emits certain forms of radiation that would be harmful to life if it reached here. The Earth has an extremely strong magnetic field which is vital for our wellbeing. It protects life from that harmful radiation from the Sun. Additionally, reaching far out into space, it protects the Earth's atmosphere and prevents it being stripped away by the Solar Wind emitted by the Sun.

The magnetic field depicted to the right has a powerful effect in protecting the Earth.

It is believed that Mars may once have had an atmosphere but, lacking an adequate magnetic field, it was stripped away by the solar wind.

Ecology

On the general wider front, with far too wide and complex a range of features and aspects to be discussed in detail here, the ecological structure of the Earth seems completely geared to the benefit and development of human life. The previous sentence should perhaps have included "when protected from human interference".

Briefly, an example of the ecological effect is that the atmosphere contains greenhouse gases like carbon dioxide that help stabilise temperature, and an ozone layer that helps protect from the Sun's radiation. These work in balance with heat absorption by the oceans and reflection by polar ice caps. Wind and ocean currents distribute heat and nutrition about the world.

The Presence of Water on Earth

There are a number of hypotheses that suggest the reason that the Earth has the amount of water present on its surface. One thing that is clear is that the heat and activity occurring at the time of formation of the Earth would have boiled off any water present. Water was therefore delivered to the Earth when the Earth had cooled sufficiently.

A currently popular theory is that the numerous comets and asteroids that were colliding with the early Earth contained large amounts of ice, and that the water on Earth was delivered thus.

Water, with its many important properties, amongst others of being able to both dissolve and transport materials, has played a complex and critical role in the development of life. It is so very important that it is worth spending time understanding it.

The chemical structure of water is almost unique, in that it is less dense in the form of ice than in its liquid form of flowing water so that ice floats on the surface of liquid water. There are many consequences flowing from this. Just one, for example, is what occurs when seas freeze over. The ice forms at the surface of the water and remains there, protecting the water below from further freezing, as long as the whole body of water doesn't freeze. The consequence of this is that the water below the ice remains as a liquid, enabling the creatures below to survive. Were ice denser than liquid water, the ice would sink to the floor of the body of water, potentially building up and creating seas that were literally massive blocks of ice.

In relation to living organisms, water is essential for their survival. The simplest way of expressing it is that virtually all the countless millions of chemical reactions that occur in living organisms occur in an aqueous environment.

Without water there would not be any life on Earth.

www.ingramcontent.com/pod-product-compliance
Lightning Source LLC
LaVergne TN
LVHW070058080426
835510LV00027B/3433